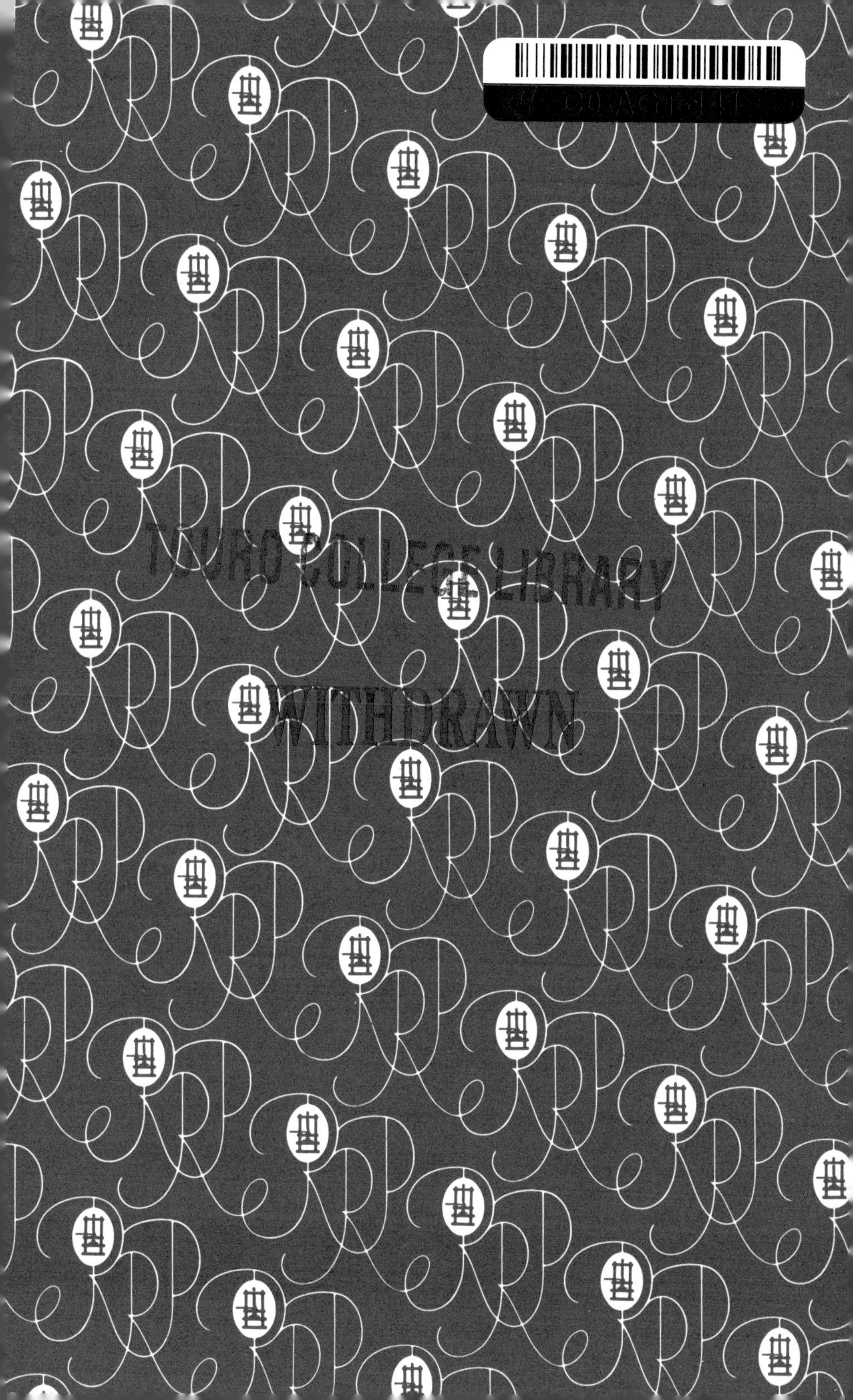

THE STUDENT EARTH SCIENTIST
EXPLORES THE EARTH
AND ITS MATERIALS

THE STUDENT EARTH SCIENTIST EXPLORES THE EARTH AND ITS MATERIALS

By

Constantine Constant

RICHARDS ROSEN PRESS, INC.
New York, New York 10010

Published in 1975 by Richards Rosen Press, Inc.
29 East 21st Street, New York, N.Y. 10010

First Edition

Library of Congress Cataloging in Publication Data

Constant, Constantine.
The student earth scientist explores the earth and its
materials.

(The Student scientist series)
1. Earth sciences. I. Title.
QE26.2.C66 550 74–23283
ISBN 0–8239–0306–0

Manufactured in the United States of America

About the Author

Constantine Constant (known as Gus to his friends) has to his credit an impressive array of publications in Earth Science, but he is a good deal more than a teacher and writer. During World War II he served as a weatherman for the United States Navy, and in the early 1950's he also was on the staff of the United States Geological Survey, a division of the Department of the Interior.

Born in Pawtucket, Rhode Island, he was educated at Hofstra University, New York University, and Hunter College, all in Greater New York, and he holds the degrees of bachelor and master. His teaching experience in various fields of science ranges from elementary school through junior and senior high school, to college. He is currently a teacher of Earth Science at Forest Hills High School, New York.

Mr. Constant's publications include *Energy Changes in the Atmosphere and Hydrosphere, Matter and Life in Space, Review Text in Earth Science,* and an *Earth Science* workbook. He has prepared experimental and regular science curricula for the New York City Public Schools, which are used all over the United States. Among his many scientific articles, a number have been translated into the French and German languages and have been published in Switzerland.

Contents

List of Illustrations

THE STUDENT EARTH SCIENTIST EXPLORES THE EARTH AND ITS MATERIALS

I

Our Planet Is a Natural Spaceship

A Place to Live Among the Stars

Have you ever looked up on a starry night and wondered—is there life out there? And if there is life out there, what does it look like? Are there other life forms that look like us? Do they live on a planet like ours? Do they have cars, homes, books, radios, television sets, games, and pets as we do? So far, no one can answer these questions.

Strange as it may seem, other peoplelike beings may even ask the same questions we ask. They may be doing the same things we do on earth. They may have farms, cities, and factories like ours. Maybe they swim, run, dance, and play ball as we do. Their natural world, also, may be much like ours. They may have rivers, lakes, oceans, mountains, and forests. Large numbers of plants and animals may live in their world. Clouds may float in an atmosphere like ours. And stars like our sun may light their planets. Of course, such planets may not exist. Of the planets circling the sun, only the earth is known to have many life forms. Some of the other planets may have some simple life forms; for example, one-celled plants may live on Mars. When we look at the life around us, we realize that our planet is special. Its atmosphere, oceans, and land formations make it possible for many kinds of life to exist.

Born Out of the Dust

Information about the origin of the earth comes from several sources. Some clues are found in the rocks that make up the earth and the moon. More clues come from studies of the sun and other stars. Other evidence is found in the dust and gases scattered in the spaces between the stars and the planets. The dust and gases are so thinly scattered that space may be considered empty; yet, if all of these particles were squeezed together, they would weigh as much as all the stars. Many scientists believe that the sun, earth, moon, and other planets were formed from such particles. They believe that dust and gas particles in space may have gathered into one gigantic ball, which heated up and broke apart to form many different sized bodies, one of which was planet earth.

Since its birth, the earth has gone through many changes. Nowadays, scientists study the earth by various methods. They drill miles below the surface. They study land forms such as mountains and volcanoes, rocks and minerals, and earthquakes. When necessary, man-made earthquakes are produced and studied. Even the earth's air and water are analyzed. From such studies, much is learned about how the earth has changed. From the evidence, scientists have put together the earth's history.

The Earth Grows Up

From evidence obtained by studying the earth, moon, sun, and stars, scientists believe the earth was born billions of years ago. Great changes took place during the earliest times.

In the beginning years, the earth was probably red-hot. It must have smoked, sputtered, and spattered for millions of years. Gradually it cooled off, and in time the first land forms and oceans appeared. Simple life forms, such as one-celled plants, probably first appeared in the oceans. As time passed, the primitive earth began to look somewhat as we know it today.

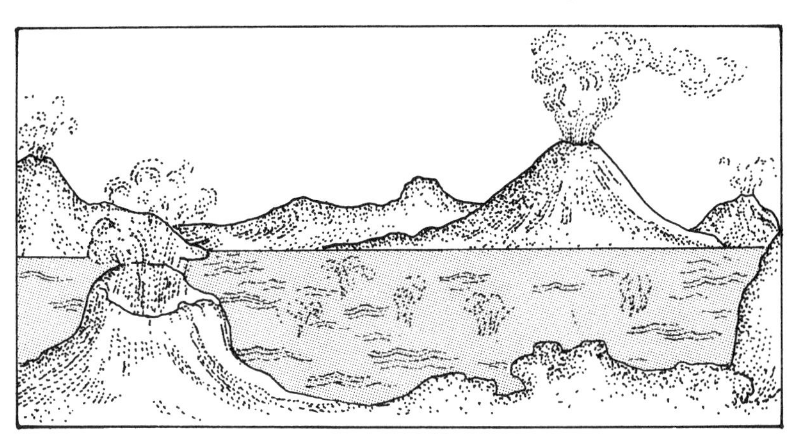

Fig. 1. In the beginning, the earth was probably red hot

In the early years, many of the rocks and minerals we find around us were formed. Many millions of years ago, most of the fuels we use today—coal and oil—were formed.

Exactly How Old Are You, Earth?

The answer to the earth's age may be found in many places. It may lie in the rocks that make up the continents. It may be found by studying the sun, moon, and stars. And some clues may come from "visitors" from space. These visitors are meteorites (shooting stars) that fall to earth from space by the millions each day. Most meteorites are as small as or smaller than grains of sand. Some weigh many tons. These particles may be the remains of a shattered planet that was born at the same time as the earth. Scientists have determined the age of these visitors from space. They estimate the typical meteorite is about 4½ billion years old. What does this tell us about the age of the earth? According to scientists, the earth is about 5 to 6 billion years old.

Compared to other planets, our planet is truly remarkable. No

Fig. 2. Meteorite on display

othcr planct in our solar system comes close to matching our rivers, oceans, mountains, glaciers, deserts, and other natural features, and our life forms. Some planets may have one or two of our natural features—but not all of them. For example, Venus and other planets have atmospheres. But their atmospheres are hot, cloudy, and made up of harmful gases. Deserts, volcanoes, and mountains are features that may be found on other planets. But no planet has life forms that abound in such places on earth.

In the chapters that follow you will find out how truly remarkable our planet is. You will learn more about its origin, and the development of its land masses, its oceans, and its atmosphere. You will see how scientists have learned much about the earth's interior. And you will see how your everyday existence depends upon the rocks and minerals in the earth's crust.

II

The Birth of a Planet

In the Beginning . . .

Can you imagine seeing planet earth being born? Most scientists agree that the birth of the earth—about 5 billion years ago—was tied in with the birth of the sun, the moon, and other planets.

Although the earth is so unlike the other planets, it seems to have formed in the same way. By studying the sun, the moon, and the planets (our neighbors in space), scientists make educated guesses as to how they formed. (As you probably know, educated guesses are called hypotheses or theories.) Important clues to the earth's formation are found in the structure and composition of our neighbors in space. From such clues, several theories about the birth of the earth have been proposed.

How Did It All Start?

According to an old theory, there was a near collision between our sun and another star. The tremendous gravity of the passing star pulled huge, flaming masses away from the sun's surface. These masses spun outward and formed a long stringlike mass of matter. The flaming masses closest to the sun were pulled back into the sun. The material farther out formed sev-

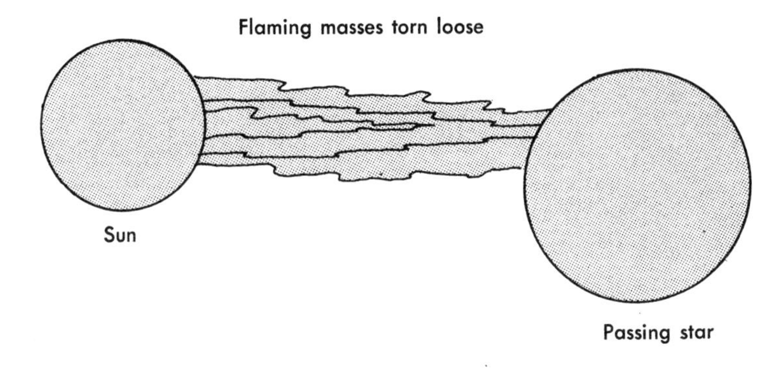

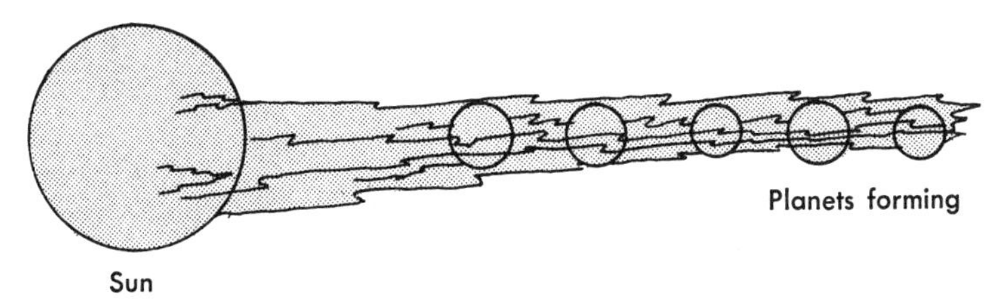

Fig. 3. The planets form

eral giant balls of molten material. In time, these molten balls cooled and formed the planets.

A Planet from Dust—A Modern Theory

Today, many scientists believe that the newly formed earth was a cold planet. They think the earth may have formed from thick clouds of dust and gas in space. These particles would attract each other and form large balls of dust and gas, which in turn would grow bigger by attracting many more smaller particles. In time, the balls would be several miles thick. The balls that grew most quickly would capture particles from smaller bodies nearby. As a final result, the original thick masses of dust and gases would be in the form of several different-sized bodies. The largest one, hundreds of thousands of miles across, became the sun. Smaller ones—thousands of miles across—became the planets. What became of the smallest bodies? Between Mars and Jupiter there is a belt made of thousands of smaller fragments that look like the parts of an exploded planet. Some of these bodies are hundreds of miles across. Because they are planet-like, they have been named planetoids. Possibly these small bodies failed to capture enough particles to become planet-sized.

The Earth, Mars, and the Moon—Once One?

If all of the planets and the smaller bodies in space were formed from the same masses of dust and gas, they should be made up of the same kinds of matter. Evidence that this is so has been found by studying these bodies by means of telescopes and rockets. Valuable information has also been obtained from shooting stars that fall to earth. Actually, shooting stars are usually small bits of matter speeding through space. When a shooting star falls to earth, it is called a meteorite. Other valuable information has been obtained from rock samples brought back from the moon.

How do these space materials help explain the origin of the earth? Many meteorites have been found to be made of iron and nickel. These two elements are believed to make up a large part of the earth's interior. Other meteorites have been found to be made of rock. Much of the earth's surface is composed of rocks or materials that come from rock. Also, rock samples brought back from the moon are quite similar to the rocks that we find at or near the earth's surface.

Such clues have led some scientists to believe that the earth, the moon, and Mars were once one huge, rapidly spinning mass. As it rotated, it broke into three parts. The largest part became the earth; the second largest part—about $1/10$ the size of the earth—became Mars; and the smallest part—about $1/80$ the size of the earth—became the moon.

Further Clues—Densities of the Planets

Important in explaining the origin of the earth is the density of the planets. Density is easily understood by comparing two marbles of the same size. If one marble is made of glass and the other of steel you will surely note that the steel ball feels heavier.

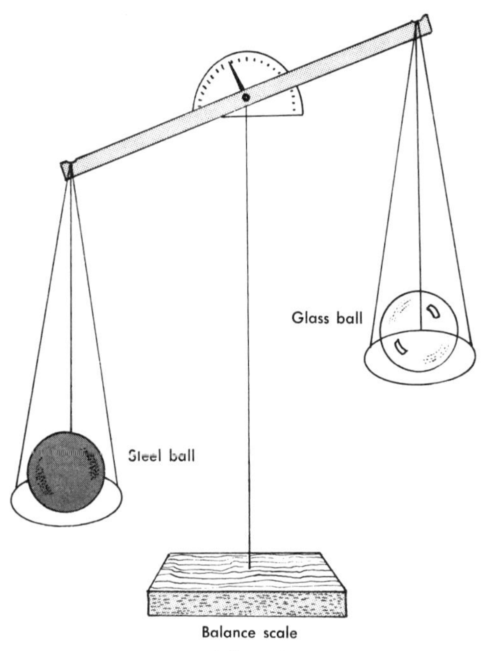

Fig. 4. Comparison of densities

Since both balls are the same size, you can state that the steel ball contains more matter. Another way of saying this is that the steel ball is more dense than the glass ball. If you placed each ball on a balance scale you would see that the steel ball tilts the scale downward. It is clear, then, that the steel ball is heavier than the glass ball.

In considering the density of the planets, you should recall that the planets formed from a mass of dust and gases along with the sun. This mass consisted of about 90 percent hydrogen gas. Hydrogen is the least dense or lightest of all substances. The planets that formed closest to the sun had just about all of their hydrogen driven off by the great heat from the sun. Planet earth was among those planets. Today, the planets that are closest to the sun are the densest.

The planets farther out from the sun were not as strongly heated by the sun's rays; thus, they remained cool. Two of these planets—Jupiter and Saturn—retained much of their original hydrogen. These planets are much less dense than Mars and earth. The other far-out planets also have densities much lower than the planets closest to the sun. See Table I.

Table I: Densities[1]

Planets	*Average Density*
Mercury	5.4
Venus	5.1
Earth	5.5
Moon	3.3
Mars	4.0
Jupiter	1.3
Saturn	0.7
Uranus	1.6
Neptune	2.2
Pluto	?
Common Materials	
Water	1.0[1]
Ice	.92
Nickel	8.9
Iron	7.5
Earth's crust	2.8
Earth's core	10.7

[1] All densities are compared to that of water. For example, consider a sample of iron having a density of 7.5. This means that a given volume of iron is 7.5 times heavier than the same volume of water.

The Moon's Density

If the moon broke away from the earth, what should the moon be made of? Studies of the moon's crust (top layer) show that it is made of rocks somewhat like those in the earth's crust. In comparing the densities of the earth and the moon, scientists have found that, overall, the density of the moon is 3.3. This is about the same as the density of the rocks that make up the crust of the earth. The overall density of the earth's crust is about 2.8. At this point, we should note that the earth's interior (core) is composed of extremely heavy elements—iron and nickel. However, the crust of the earth is composed of rocks that are much lighter. Actual calculations of the overall densities of the moon and the earth indicate that the earth is about one and a half times denser than the moon. Add to this the fact that Mars (the planet closest to the earth) is a little less dense than the earth, and we can see how the earth, Mars, and the moon may have once been one giant planet that broke apart. Some scientists believe that the moon may have been torn loose from the region now occupied by the Pacific Ocean.

Whatever explanations come up in the future, it is doubtful that we will ever fully explain the origin of the earth. The puzzle, however, impels people to probe the earth and its neighbors in space for more clues to the earth's origin. Many valuable clues are to be found in the land masses of the earth, its oceans, and its atmosphere.

III

Shaping the Earth

Our Newborn Earth

Picture the earth shortly after it was formed—possibly bubbling, flaming masses at the surface and steaming clouds of gases swirling about above. The earth may have looked like this for millions of years. In time, some of the gases may have cooled and fallen as rain. The rain may have fallen for many thousands of years. Eventually, the earth may have been completely covered by water.

Keeping the above picture in mind, let us look back in time and try to picture the formation of the earth's continents, its oceans, and its atmosphere.

The Earth's Land Masses

As you know, the continents are the major land masses upon which man dwells. If the theory that Mars and the moon came from the earth is true, the continents may be those parts of the earth's surface that did not break loose. Before this time, you recall, the entire earth may have been covered by water.

When Mars and the moon were torn loose, huge gaping holes were left in the earth's surface. These holes, which were thousands of feet deep, quickly filled with water, uncovering large masses of land. These land masses that stood high and dry were the newly formed continents. Then, as today, the continents made up about one fourth of the earth's surface.

The Continents Move Apart

Many scientists agree that the continents were once one huge land mass. About 150 million years ago, this supercontinent began to break up. In some way, the continents slowly drifted apart. Their locations today indicate that they have moved thousands of miles from their original positions. Scientists explain this movement by the theory of continental drift.

The Theory of Continental Drift

According to the theory of continental drift, strong, slow-moving currents of molten materials lie beneath the earth's crust. In some places, these currents lift the crust; in other places, they drag the crust downward. The up-lifting and down-dragging of these currents cause the crust to buckle and break. As a result of these movements, long thin breaks appear in the ocean floor. Each time the breaks occur they fill with materials from below. Once the molten materials harden, the continents cannot come back together. Thus, each break causes the ocean floor to spread apart a little bit.

By means of simple arithmetic, we can see how these little breaks caused the continents to be pushed thousands of miles apart. Assume that every 10 years the ocean floor moved apart 1 foot. Take 150,000,000 years, the time when the supercontinent broke apart, and multiply by 1 foot—the answer is 150,000,000 feet. Then $\frac{150{,}000{,}000}{10} \div 5{,}280$ (feet in 1 mile) = 2,841 miles, the distance the continents could have moved apart.

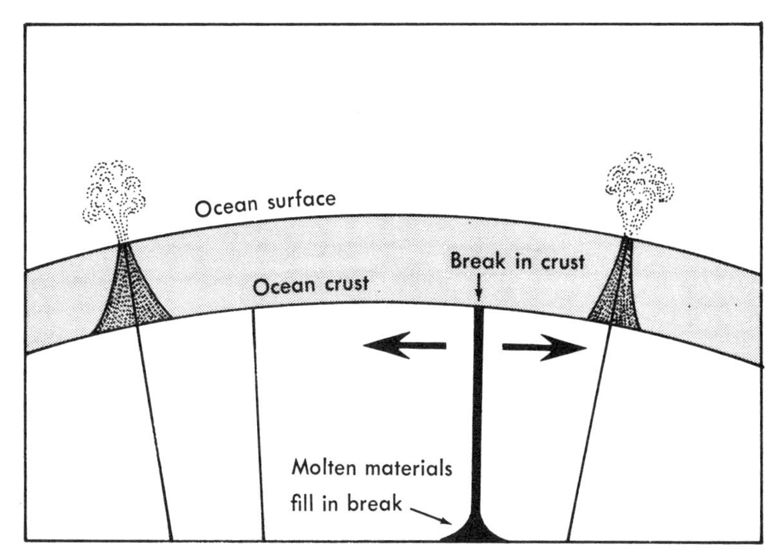

Fig. 5. Continental drift

The Age of the Ocean Floor

Studies of the rocks that make up islands show that islands are younger than the continents. Scientists believe the islands may have formed as the spreading ocean floor caused the continents to drift apart. Accordingly, the islands that formed first would now be farthest away from the center of the ocean. Measurements do indicate that the oldest islands are usually farthest from the center of the ocean, giving evidence that the continents may have drifted apart.

More Evidence of Continental Drift

Another way to tell that the continents were once joined together is to study the shapes of their shorelines. As you can see in Fig. 6, the coast of Africa fits like a jigsaw-puzzle piece into the coasts of North and South America. More and better evidence

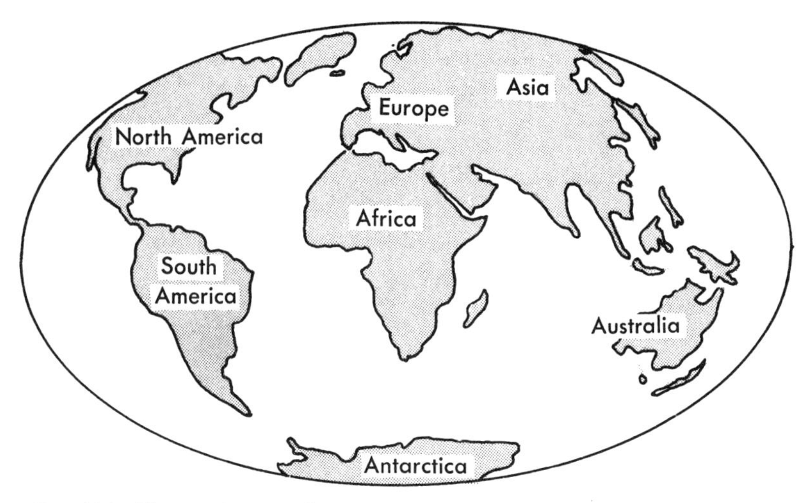

Fig. 6(a). The continents today

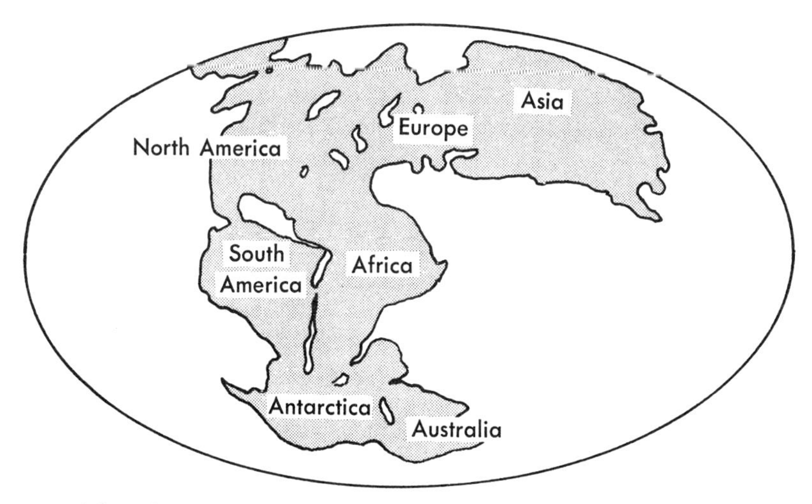

Fig. 6(b). The continents 150,000,000 years ago

however, is found within the rocks that make up these coastlines. The evidence is in the form of lines of magnetism "frozen" into the rocks on either side of the Atlantic Ocean. These lines were formed by the earth's magnetic field. You can produce similar lines by placing a magnet beneath a thin piece of cardboard. Sprinkle iron filings on the cardboard, then gently tap the cardboard. The iron filings will line up. If the iron filings were then glued into place, you would have a permanent record of the magnetic field around a magnet. Similar magnetic lines formed as molten rocks in the newly formed supercontinent cooled. To this day, these magnetic lines can be found in the rocks found along the shores of the Atlantic Ocean. The magnetic lines frozen into the rocks along the shorelines of North and South America and Africa are similar. Since these magnetic lines form similar patterns, they are strong evidence that they were formed in the same place—where the continents were once together.

The Oceans Are Born

Actually, what we call the oceans is really one huge ocean that covers about 71 percent of the earth's surface. This ocean, you recall, may have been shaped when large masses were torn loose in the formation of the moon and Mars. Before this time, it is believed that the earth was completely covered with water. Where did the water come from?

Water from Rocks

Experiments have shown that at high pressures and high temperatures water mixes with molten rock. Thus, as molten rocks cooled at or near the surface, some of the water was released as steam. This steam may have surrounded the earth as a thick cloudlike blanket. Eventually, the clouds cooled and the steam changed to liquid water. The water then fell to earth as rain. In time, the entire earth was flooded by water. Since rainwater is quite pure, the original ocean consisted of fresh water. How, then, did the oceans get salty?

The Salty Ocean

If you have ever gone swimming in the ocean, you surely noticed that the water was salty. The salts in the ocean were once part of the land. As rainwater fell on the land, some of the salts in the rocks were washed out. Over millions of years, these dissolved salts were carried by rivers into the ocean.

Today, every 100 pounds of ocean water contains about 3½ pounds of salts. The largest part of these salts is table salt—the same salt you put on a hamburger and French fries. The actual salt content of the ocean is shown below.

Table II: Salts in the Oceans

Salt	*Formula*	*Percent (approx.)*	*Common Name*
Sodium chloride	$NaCl$	78	Table salt
Magnesium chloride	$MgCl_2$	11	
Magnesium sulfate	$MgSO_4$	5	Epsom salts
Calcium sulfate	$CaSO_4$	4	
Potassium sulfate	K_2SO_4	2	
Calcium carbonate	$CaCO_3$	1	Limestone
Magnesium bromide	$MgBr_2$	1	

To get the best picture of how much salt is in the oceans, imagine that all of the water were evaporated and the salts remained on the ocean floor. The salts would form a layer approximately 175 feet thick over the entire ocean floor.

The Ocean Floor Is Flat

The shape of the ocean bottom is best described as being mostly flat with occasional islands and mountains rising upward from the floor. Deep cuts (trenches) in the bottom of the ocean make up the deepest parts of the ocean. Whereas the average depth of the oceans is 2½ miles, these trenches plunge downward nearly 7 miles.

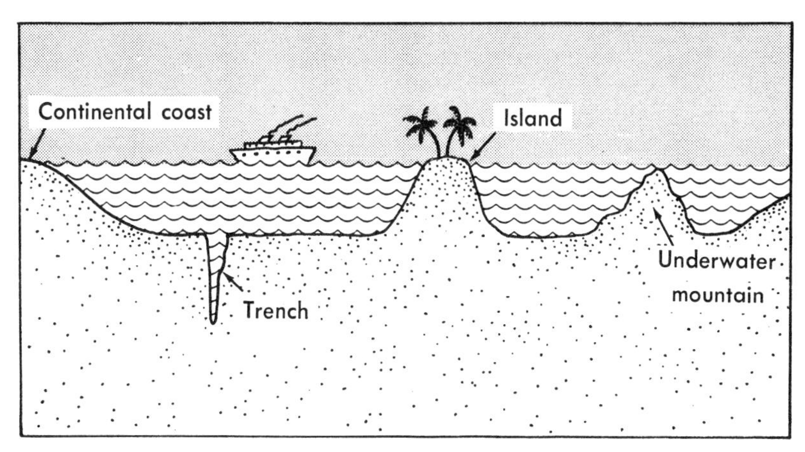

Fig. 7. The ocean floor

Oxygen: The Breath of Life

We live in an ocean of air, which extends upward for thousands of miles. This ocean of air is called the atmosphere. Over half of the air, by weight, lies below a height of about 3½ miles.

An important part of the air is oxygen. If you hold your breath for a short period of time, the need for oxygen is readily felt. This is notable, because the earth's first atmosphere was probably poisonous to life. There was no oxygen for breathing. Some of the poisonous gases were ammonia (NH_3) and other compounds containing elements such as sulfur and chlorine. At a later stage, our atmosphere may have consisted of gases similar to the gases given off by volcanoes today. Such gases are water vapor (H_2O), carbon dioxide (CO_2), and sulfur compounds (H_2S, SO_2).

Oxygen Enters the Air

The atmosphere changed with the appearance of plants on the earth—probably well over a billion years ago. Plants breathed in carbon dioxide, and in their food-making process released oxygen as a waste product. As more plants appeared on the earth, more oxygen was released into the air. For the past billion years, our atmosphere has changed very little. The composition of this atmosphere today is shown in the table below.

Table III: Gases in the Atmosphere

Gas	*Symbol*	*Percent by Volume*
Nitrogen	N	78.1
Oxygen	O	20.9
Argon	A	.9
Carbon dioxide	CO_2	.03
Neon	Ne	.00002
Helium	He	.000005
Hydrogen	H	.0000005
Other	—	less than .003

IV

Measuring the Earth

The Shape We Are In

All of us know that Christopher Columbus, Ferdinand Magellan, and other explorers helped to prove that the world is round. Surprisingly, however, a group known as the Flat Earth Society meets annually and claims that the earth is really flat. No matter what arguments others give to prove that the world is shaped like a ball, by their own distorted logic they twist words and ideas and end up with a flat earth. Despite their claims (they must be joking), there are several ways in which we can prove the shape of the earth.

The Horizon Is Curved

Have you ever stood at the shore of a large lake or the ocean? If you have, you may have noticed that the horizon gently curved from left to right. What you did not see was the earth's surface straight ahead of you curving gently downward.

If you ever watched a ship sail toward the horizon, you would have noticed the effect of the earth's curvature. What would you have observed? First, the lower parts of the ship would disappear from view, then the upper parts would disappear. Finally, the ship would be out of sight.

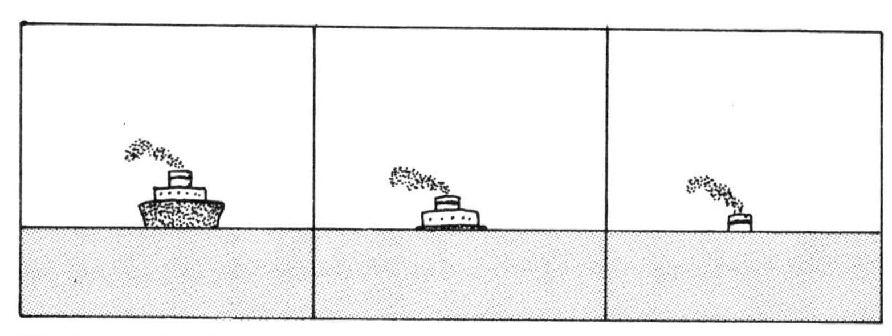

Fig. 8 (a). The ship "disappears"

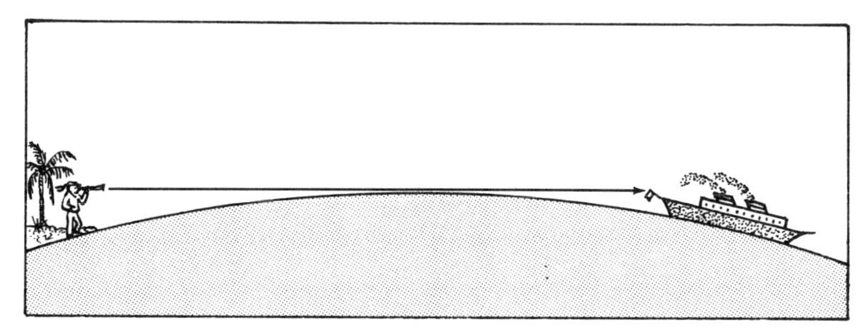

Fig. 8(b). This illustration shows what really happens

Actual measurements show that the earth's surface curves very gently—about 8 inches per mile. Thus you can see why a ship slowly disappears from view. Each mile that it sails away from you, it drops about 8 inches out of your sight.

The Earth's Curved Shadow

Here is a challenge for the Flat Earth Society. At times the earth's shadow falls upon the moon. This effect we know as an eclipse. At such times, people on the earth can see the earth's curved shadow move slowly across the face of the moon. From your own experience, you know that the shadow cast by a ball is curved. Could a cylinder or a pie plate produce a curved shadow that is always round? Both could! However, both would always

Fig. 9. Round shadows

have to be lined up in a special way with the source of light. If you turned them about they would form several different-shaped shadows. A ball, of course, would always cast a round shadow no matter how you turned it about.

Exactly What Is the Shape of the Earth?

Suppose you were an astronaut circling above the moon. What would the earth look like from this very great distance in space? You would probably describe the earth as a huge round globe. Actual photographs taken from space show the earth to be a globelike mass. Large sections of this globe are seen to be covered by swirls of clouds. Much of the earth's surface that can be seen beneath the cloud cover are the ocean waters, which appear as blue and green patterns. In addition, sections of the continents show up as patches of green, yellow, and brown.

A good way to examine the earth's shape is to shrink it to the size of a handball. Assume that everything on the earth shrinks accordingly. Then the handball-sized earth would feel as smooth as a billiard ball. As seen this way, the earth would also appear perfectly round.

Actual measurements show that the earth is slightly out of shape. For example, the earth's polar circumference is 24,860 miles. The equatorial circumference of 24,902 miles is slightly larger. These measurements tell us that the earth bulges slightly around the equator. The bulge is caused by the earth's spinning motion. In your own experience you may have seen a girl ice skater spinning in place. The spinning motion caused her skirt to be thrown outward. Scientists describe this outward flinging of a whirling object as the centrifugal effect. It acts on the entire crust of the earth. The overall result is to flatten the earth at the poles and to make it bulge at the equator. We can simplify this and state that the earth is a sphere that is slightly flattened at the poles. Scientists use the term oblate spheroid to describe this shape.

Using Sunlight to Measure the Size of the Earth

About two thousand years ago Eratosthenes, a Greek geographer, measured the circumference of the earth very accurately. He assumed that the earth was shaped like a sphere. Since he could not walk, sail, or fly around the earth, he could not measure it by any of these methods. He found his solution in rays of sunlight falling on the earth. He noted that at noon the sun was directly overhead at Syene, Egypt. Thus, the sun cast no shadows. At the same time, at Alexandria, 489 miles away, the sun cast short shadows. By his measurements, he found the sun was about 7.2 degrees away from the overhead position at

Alexandria. Now you can see how he solved the problem of determining the earth's circumference. The solution came from the angle of sunlight between Syene and Alexandria—7.2 degrees. This angle also represents 7.2 degrees of arc on the earth's surface.

The distance between Syene and Alexandria is 489 miles. This distance represents 7.2 degrees of a full circle. How many miles are there in the full circle that goes around the earth?

There are 360 degrees in a circle. Then 7.2 degrees divided by 360 degrees = $\frac{1}{50}$. Since 7.2 degrees is $\frac{1}{50}$ of a circle, and it also represents 489 miles, we can say that the circumference of the earth equals 489 times 50 = 24,450 miles.

By their method, Eratosthenes measured the circumference of the earth. Surprisingly, he was very close to today's measurement of 24,902 miles (at the equator).

Using Satellites to Measure the Earth

As you know, the United States has placed many satellites in orbit around the earth. These satellites are often tracked by means of radar as they circle the earth. It is simple, then, to see how a satellite circling the earth can be used to measure the size of the earth. Basically, all a scientist has to know is the speed of an orbiting satellite. Next he determines how long it takes the satellite to make one complete orbit around the earth. Then, by simple arithmetic, he finds out how many miles the satellite traveled around the earth. For example, assume a satellite is circling the earth at 17,000 miles per hour and it makes one complete orbit in 1½ hours. (For this problem, assume that the satellite is traveling in a close-to-perfect circular orbit.)

Then Rate $\times$ Time = Distance traveled
17,000 $\times$ 1.5 = 25,500 miles

Since the satellite is circling above the earth, a small adjustment has to be made, depending on the altitude of the orbiting satellite. Since the satellite's exact altitude is known, this figure is subtracted from the radius of the circle that it traces around the earth. The new radius is the earth's radius. Using this information, we can calculate the earth's exact circumference.

For example:

Circumference of satellite's orbit $= 2\pi r$

$= 2 \times \pi \times 4264$

Circumference of earth $= 2 \times \pi \times [4264 - 300$ (satellite's altitude)]

$$C = 2 \times 3.14 \times 3964$$
$$C = 24{,}894 \text{ miles}$$

Satellites can also be used to measure the shape of the earth. This is done by using a satellite's altitude as a kind of measuring stick. As a satellite circles the earth, its exact height is measured from many points beneath it on the earth's surface. These points, when connected, show the shape of the earth beneath the path of the satellite. Thousands of such measurements, along different paths, show the shape of the earth very accurately. This shape, you recall, is known as an oblate spheroid.

Can the Earth Be Weighed?

Suppose you were handed a rock and told to weigh it? You would look for a scale to place it on, and within a short time you would know what the rock weighs. That is simple enough. But suppose someone asked you to weigh an ocean liner? Or, suppose you were asked to weigh the earth? Although both tasks seem impossible, we can calculate the weight of both. To calculate the weight of an ocean liner is quite simple. Without going into great detail, the weight of the ocean liner is equal to the weight of the volume of water it pushes aside to remain afloat. As you can readily see, however, weighing the earth presents a far more difficult problem. There is nothing but empty space around the earth. Thus, there is nothing for the earth to push aside.

The problem of determining the weight of the earth was solved by the German physicist Philipp von Jolly. His method was simple and quite accurate. We should note, however, that von Jolly did not actually weigh the earth. He knew that the weight of an object is measured by noting the pull of gravity on its mass. For example, your weight depends on your mass, that is, the amount of matter that makes up your body. When you stand on a scale, the pull of gravity acts on your mass. The effect of gravity on your mass is shown on the scale as pounds. Furthermore, the amount of mass beneath you affects your weight. Where, then, would you weigh less, on the moon or on the earth? The answer, of course, is on the moon, for the moon's mass is much smaller than the earth's. Although your weight changes as you move from one place to another, your mass always remains the same.

Since von Jolly could not weigh the earth, he determined its mass by using a large balance scale. First, he placed a large mass on one side of the scale, then he balanced the scale by placing an equal amount of mass on the other side. At this point the pull of gravity had the same effect on each side of the scale. The scale was balanced. Then he placed tons of lead beneath one side of the scale. What do you think happened? The side of the scale with the mass of lead beneath it was pulled downward. It became heavier, because the amount of mass beneath it had increased. Next, von Jolly added known masses to the other side of the scale to balance the scale once more. The calculations that followed were quite complicated. But, once the earth's mass was known it could be stated as weight. According to von Jolly's calculations, the mass of the earth is about 5,980,000,000,000,000,000,000,000 kilograms, or close to 7,000,000,000,000,000,000,000 U.S. tons.

Once we have determined the earth's mass, we can also determine its density. Density is determined by dividing the earth's mass by its volume ($D = m/v$). The result of this division is about 5.5. This number, you recall, tells us that overall the earth is about 5.5 times heavier than an equal amount of water. If you refer back to the *Table of Densities* on page 11, you can see how the earth's overall density compares to that of other planets.

V

The Earth's Interior

The Mystery of the Earth's "Inner Space"

"Out of sight, out of mind." Have you ever heard that saying? According to the saying, what we cannot see within the earth need not concern us. Observations and studies indicate, however, that the earth's interior is quite active. Forces within the earth buckle and break the crust. The continents have been pushed great distances apart. Although the crust is solid, molten materials circulate beneath it. Such activities have aroused the curiosity of many scientists. And students ask numerous questions about the earth's interior: Can you dig a hole all the way through the earth? What is it like deep inside the earth? Are there underground lakes? How hot is it? What is it made of? Is the earth solid all the way through?

The first question is easy to answer, for man has drilled thousands of holes into the earth by means of drills. The deepest of these holes penetrated about 5 miles. Yet this hole is not even a pinprick compared to the total size of the earth.

Drilling Through the Crust

Much of our knowledge about the earth's crust comes from

drill holes (also called bore holes). The materials that are brought up from such holes tell us much about the nature of the earth's crust.

Holes are drilled for several reasons. For example, drills are used to search for oil, water, gas, and other mineral deposits. Holes are also drilled to study underground formations. Such information is especially useful for builders of dams, bridges, and other large structures. Holes are also drilled purely for the gathering of scientific data. The information gathered from all of these bore holes gives us a good picture of the composition and structure of the crust.

The Earth's Crust is Mostly Oxygen

You recall that oxygen is the second most abundant gas in the atmosphere. Would you believe that oxygen makes up about 94 percent of the earth's crust? If we consider the total volume of the earth's crust, nearly 94 percent of that volume is oxygen. Of course, you cannot breathe that oxygen. It is chemically combined (tied up) with other elements so that it is not free. In such combinations it makes up the rocks and minerals in the earth's crust. Some elements to which oxygen is tied are aluminum, silicon, iron, and calcium. These are among the most abundant elements in the earth's crust.

Another way of indicating the amount of oxygen in the crust is by weight. When we measure by weight, the result may not be as startling as when we measure by volume. By weight, oxygen makes up nearly 47 percent of the earth's crust.

You may wonder why gold, silver, copper, lead, and other common elements have not been mentioned as abundant elements. Although these elements are common to us, they are rare in the crust. As you can see in Table IV, the five elements listed make up over 90 percent of the earth's crust. All of the other elements make up less than 10 percent of the earth's crust. Of

Table IV: Some of the Most Abundant Elements in the Earth's Crust

Element	*Symbol*	*Percent (approx.) Volume*	*Weight*
Oxygen	O	94.0	47
Silicon	Si	1.0	28
Aluminum	Al	.5	8
Iron	Fe	.4	5
Calcium	Ca	1.0	4

these, copper makes up about 0.0005 percent, and gold makes up about 0.0000007 percent.

Continental Crust Versus Oceanic Crust

Often the crust is hidden from our view by soil, sand, sidewalks, highways, and buildings. At times, the crust is not covered and we recognize it as solid rock. This solid rock may be seen right in your own neighborhood. It may have been the huge mass of rock you sat on in a park. Maybe you saw large sections of it blasted loose by construction men, and later you saw a building erected on the same spot. Possibly, it was a large rock you tried to dig up in your own backyard. The more you dug, the more rock you may have seen. If you could not uncover the entire rock, it may have been an exposed part of the crust. Such exposures are called outcrops.

If we could scrape away all the loose materials and other coverings of the earth, we would see the crust as one continuous mass of solid rock.

The rocks that make up the crust of the continents are largely light-colored masses. The crust of the ocean basins is composed largely of dark-colored rocks. The rocks are different in color because they are made of different materials. Very large masses of the continental rocks are granite type rocks, which contain large amounts of silicon and aluminum. The rocks of the ocean floor are basalt type rocks, which are rich in silicon, magnesium, and iron.

On the average, the rocks that make up the continents are less dense than those that make up the ocean floor. The average density of the continental rocks is about 2.7, whereas the rocks of the ocean floor have an average density of about 3.0. Overall then, the continental rocks are lighter in weight than those of the ocean floor. Accordingly, it appears that the continents may be floating on top of heavier rocks. The heavier rocks may be those of the ocean floor extending beneath the continents. Or, the continents may be floating upon rocks similar to those that make up the ocean floor.

Are Earthquakes Useful?

As you probably know, earthquakes are often deadly. When the earth shakes violently, buildings tumble, dams break, bridges snap, and many lives are often lost. Most earthquakes, however, cause little or no destruction.

The shock waves sent through the earth by quakes are measured by means of special instruments called seismographs. The readings obtained from such instruments tell scientists what they cannot learn by means of drills. In fact, scientists produce man-made earthquakes by setting off high explosives in holes in the ground. The shock waves produced go far beyond the depths reached by drills. You recall that drill holes only go about 5 miles into the crust. By comparison, quake shock waves travel thousands of miles.

As a result of the study of earthquake waves, scientists have decided that the earth is not solid all the way through. It appears that the earth is composed of four different zones: the crust, the mantle, the outer core, and the inner core. These zones vary greatly in thickness and range from the solid state to the liquid state.

The Crust Is "Paper-thin"

In 1909 the Yugoslav scientist Andrija Mohorivicic determined the thickness of the earth's crust. He accomplished this by studying records of earthquake waves. Such records showed that several miles below the surface earthquake waves were bent from their paths. Mohorivicic reasoned that the crust ended at

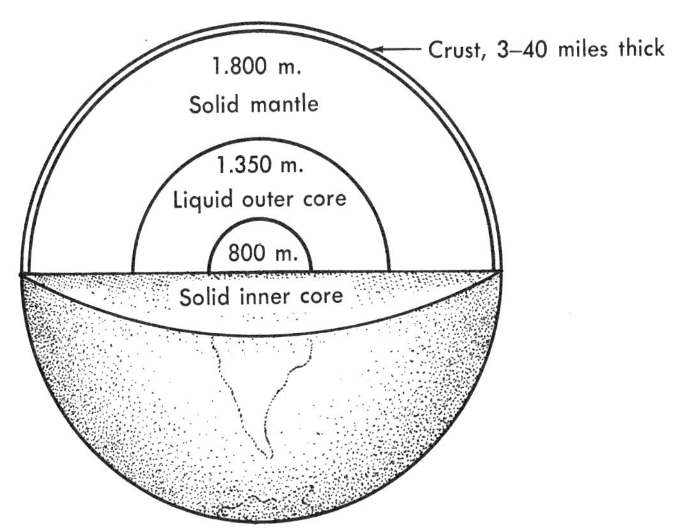

Fig. 10. A cross-section of the earth, from crust to solid core

this point. This region has been named the Mohorivicic discontinuity, or simply the Moho.

Scientists reasoned that the waves bend because the material beneath the crust is different in certain ways. According to their measurements, the material beneath the crust is more rigid (stiffer) and more dense than the crust. Their measurements also indicate that the crust of the continents is thicker than the crust beneath the oceans. On the average, shock waves bend 18 to

25 miles below the surface of the continents. Below the surface of the oceans, bending takes place at depths of about 6 to 8 miles. Accordingly, the average thickness of the continents is 18 to 25 miles, and the average thickness of the ocean floor is 6 to 8 miles. The thickest part of the crust has been found beneath the continents—about 40 miles thick. The thinnest part of the crust is beneath the oceans—3 miles thick. In comparison to the total thickness of the earth—8000 miles—the crust is paper-thin.

You recall that drills have penetrated to a depth of about 5 miles. Would it not seem easy, then, to drill downward and reach the zone beneath the crust? Where would you drill? Where the crust is the thinnest, of course—beneath the oceans.

Drilling a Hole in the Ocean

Scientists can only make educated guesses as to the composition of the material beneath the crust. Some evidence appears to come from erupting volcanoes. It is believed that the lava (molten rock) that pours out of volcanoes may come from the mantle, the layer that is beneath the crust. To find out if the mantle contains similar materials, samples are needed.

In the 1960's scientists attempted to drill a hole through the crust of the ocean floor. They called their attempt Project Mohole. A special ship was outfitted for drilling in deep water. In test drillings, samples of basalt were removed from the crust beneath the oceans. Before scientists could drill all the way through the crust, money for the project ran out. Today, new ventures for drilling through the ocean floor are being planned and undertaken. Hopefully, a drill will break through the crust before the end of the 1970's. When this does happen, scientists will be able to study samples of materials brought up from the mantle. Only then will they really know what the mantle is composed of.

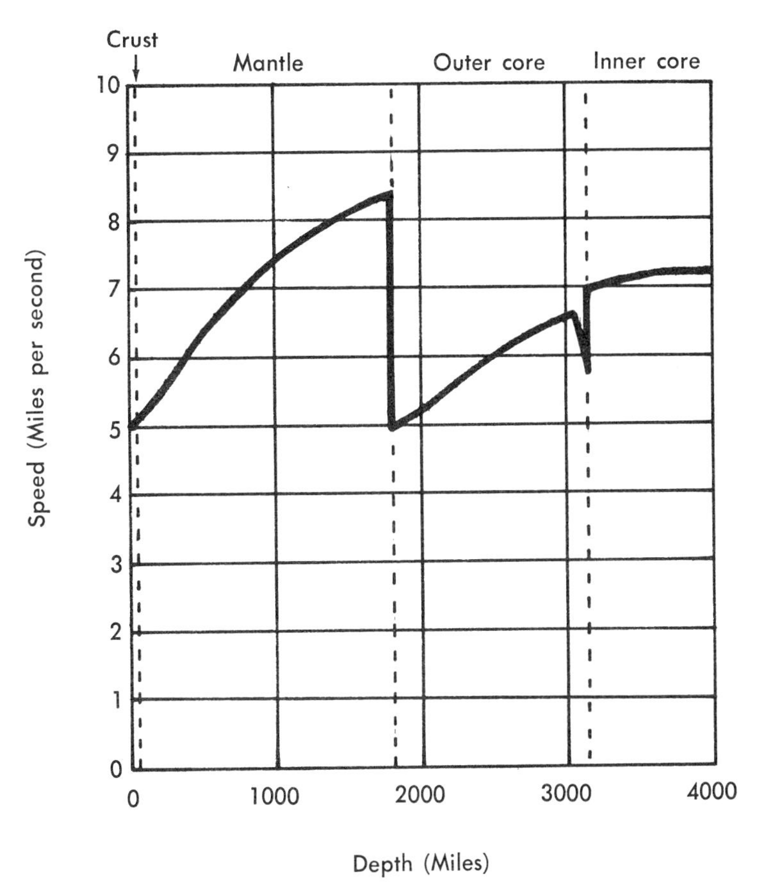

Fig. 11. The speed of shock waves passing through the earth

Beneath the Crust

We have seen how scientists study the earth's crust. We know that much of our knowledge of the deeper parts of the crust comes from the study of earthquake waves. We are also aware that much of the knowledge we have about the earth's inner zones comes from such waves. In the study of shock waves traveling through the earth, changes in speed and direction are noted. In Fig. 11 you can see how speeds change as waves travel through the earth. Where do sudden changes occur? As you can see, such changes take place between the crust and the mantle, between the mantle and the outer core, and between the outer core and the inner core. Using such data, and their knowledge of chemistry and physics, scientists can make intelligent guesses about the nature of the earth's interior.

You recall that the crust is composed mostly (by weight) of the elements oxygen and silicon. Scientists believe that the zones beneath the crust have compositions that are quite different. According to them, the mantle is rich in iron. The outer core and the inner core are rich in iron and nickel.

VI

Topographic Maps Show the Shape of the Land

A Useful Tool

Did you ever look at a hill or a cliff and wonder how high it was? Suppose you went on a hike in a strange place, would you not want to know what lies ahead? The location and size of cliffs, deep holes, rivers, and falls would be very helpful. A map that gives such information, and much more, is the topographic map.

Why Should We Learn To Read Topographic Maps?

The topographic map is a handy tool for campers, hikers, engineers, and geologists. It is especially useful to students to show features of the land being studied. Places that you may never be able to visit can be brought right into the classroom. You can study features such as volcanoes, glaciers, mountains, plateaus, rivers, lakes, and canyons without ever leaving your seat. And, if you do wish to visit and study these and other land features, a topographic map should be taken along to answer questions you may have, about height, steepness, distance, etc.

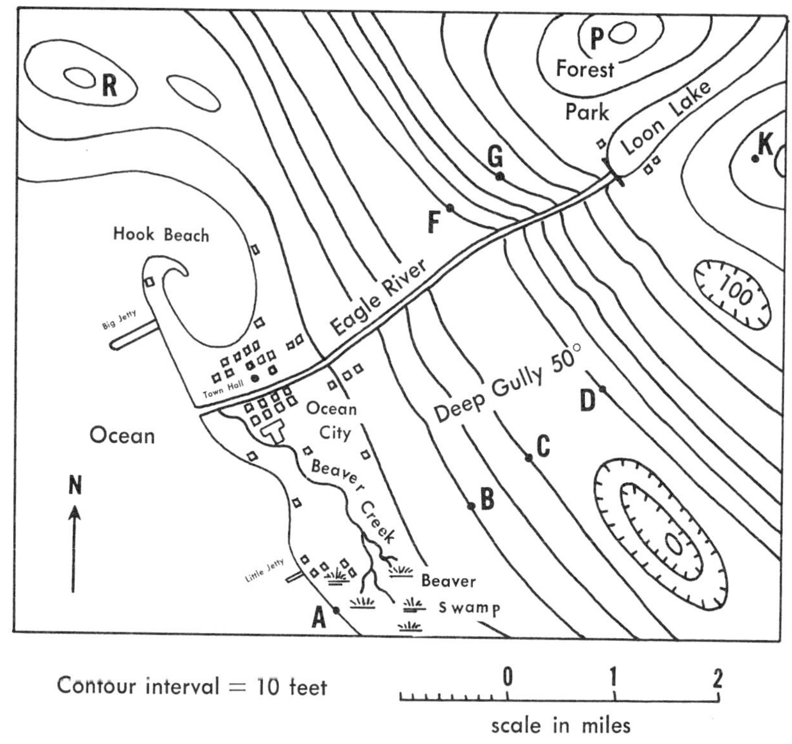

Fig. 12. A topographic map

How Is a Topographic Map Made?

Have you ever seen an aerial photograph? This is a photograph of the ground taken from an airplane. Aerial photographs show the location and size of rivers, lakes, woodlands, farms, roads, and buildings. Can you imagine what an aerial photograph of your neighborhood would look like? If you had such a photograph, you could use a marking pen to outline buildings, roads, and other features, and from such an outline you could produce a simple map. If you wished to produce a topographic map, you would have to show the height of the land. The height of the land can be determined by means of many measurements. These measurements would be placed on the outlined photograph, and then lines would be drawn connecting all places having the same height. On topographic maps, these lines are called contour lines.

What Can We Learn from Contour Lines?

Knowing how to read contour lines is important in the use of topographic maps. Contour lines tell us many things, such as (1) the height of the land; (2) the steepness of the land; (3) the shape of land forms; and (4) the direction of flow of rivers.

How High Is It?

We read the height of the land by studying the numbers on contour lines. To help us, the mapmaker states the contour interval at the bottom of the map. For example, if the contour interval equals 10 feet, the space between any two contour lines equals a height of 10 feet. As you can see in Fig. 12, the first contour line—labeled A—is at sea level or 0 feet. Now, count the spaces from the 0 to the line numbered 50. Since each space equals a height of 10 feet, you know that from 0 to 50 equals a height of 50 feet. Thus, if you decide to walk from the Little Jetty, at the 0 line, to the 50-foot line, you know that you have a 50-foot climb ahead of you. When you reach the 50-foot line you will know that you are 50 feet above the 0 line.

The contour interval used depends on the steepness of the land. In relatively flat places, contour intervals of 10 feet, 5 feet, and 1 foot are commonly used. In steep or mountainous places, intervals of 50 and 100 feet are commonly used. By simple counting you can determine the heights of the points shown on the drawings. Thus, point A is at a height of 5 feet; point B is at a height of 100 feet; and point C is at a height of 500 feet. How high is point D?

What Does a Hole in the Ground Look Like?

To show a hole in the ground, contour lines have comblike lines coming out of them. These lines point to the bottom of the hole. In general, the depth of a hole is measured by counting the number of comblike lines in the hole. Thus, if there are four lines and the contour interval is 10 feet, the hole is about 30 feet deep.

How Steep Is It?

The steepness of the land is shown by the spaces between contour lines. The steeper the land gets, the closer together are the contour lines. The flatter the land gets, the farther apart are the contour lines. As you can see in Fig. 12, the land is steepest between letters F and G. Relatively flat places are located between letters C and D and letters A and B. How steep do you think it is at letter K?

What Is Its Shape?

The shape of land forms is easily seen, because contour lines take the shape of the land. For example, if a hill is round, the contour lines look like a bullseye (see letter P in Fig. 12). If a hill is oval, the contour lines are football shaped (see letter R).

If the land has gulleys in it, these too can be seen; wherever contour lines cross gulleys, the lines bend uphill (see Deep Gulley, Fig. 12).

In What Direction Does It Flow?

As you know, rivers flow in valleys. River valleys are like gulleys. When contour lines cross river valleys they bend as they do with gulleys. Thus, contour lines bend uphill or upstream when they cross river valleys. On maps, the contour lines look like V's or U's pointing upstream. Notice Eagle River in Fig. 12. The contour lines crossing the river point toward Forest Park. Upstream, then, is toward the park. Now you know that the river flows downhill toward Ocean City.

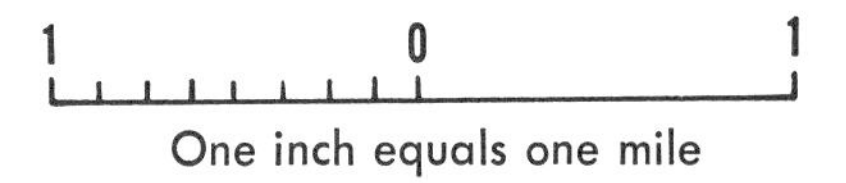

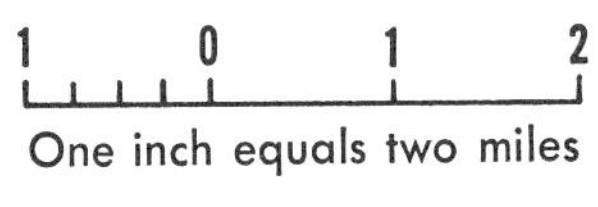

Fig. 13. Map scales

How Can We Measure Distances?

In your own neighborhood you may wonder how far it is from your home to your school. In some neighborhoods, you can count the blocks. If you know the length of each block, you can determine how far it is to your school. You can find out the length of blocks on almost any local map. One kind of local map, a road map, can be obtained from most gas stations. On a road map you will find a scale that looks somewhat like a small ruler. Examples of such scales are shown in Fig. 13. Notice that each scale is different. On the top scale 1 inch equals 1 mile, and on the bottom scale 1 inch equals 2 miles. By using the scale on your map, you can measure the distance to your school. The same type of measurements can be made on topographic maps. And the mapmaker gives you more help by telling you what the scale is in words. Beneath the rulerlike scale he may state that 1 inch equals 1 mile. On such a map, two points that are 10 inches apart are really 10 miles apart. If the scale reads 1 inch equals 2 miles, the two points are 20 miles apart ($2 \times 10 = 20$ mi.).

Look at Fig. 12. What is the scale? Measure the distance between Town Hall in Dodge City and Loon Lake that feeds Eagle River. How far is it from Town Hall to Loon Lake? Your answer should be 4 miles ($1 \times 4 = 4$). Now measure the length of Big Jetty. Your answer should be ½ mile.

Try one more measurement. Suppose you wanted to hike from Town Hall to Beaver Swamp—how many miles is it in one direction? The answer is 2 miles.

In What Direction Is It?

Mapmakers help us find directions by showing where north is. On topographic maps they do this by means of small arrowlike lines that point north. Look at the arrowlike line in Fig. 12. Note the N near the top of the arrow; north is toward the top of the map, and south is toward the bottom. Do you know in which direction east is? When facing north, east is to your right.

Often you may want to travel to places that are not exactly north, or south, or east, or west. Then you have to estimate the direction in which you want to go. This is rather simple. Just look at Fig. 12. Suppose you had to follow an arrow drawn halfway between north and east. You would travel northeast. If your path takes you halfway between south and east, you will travel southeast. What would your path be halfway between north and west?

Look at Fig. 12. Eagle River flows toward Dodge City. In what direction does Eagle River flow? Southwest is correct. If you hike from Ocean City to Beaver Swamp, in what direction must you go?

Remember, on most maps north is at the top of the map. To be sure, however, look for the north indicator supplied by the mapmaker.

VII

Learning to Recognize Minerals

Some Minerals Are Single Elements

We discussed in Chapter 2 how scientists think the earth may have formed. The same substances that made up the original earth are still present today. These substances are the elements.

So far, scientists have discovered 88 natural elements on the earth. In addition, 17 elements have been produced artificially. Any substance that exists in nature contains one or more of these natural elements. Even germs are composed of elements.

You recall that oxygen, silicon, iron, aluminum, and calcium are elements that make up much of the earth's crust. Although each of these is a single element, they are not free. These elements are chemically combined with other elements in the crust. In the combined form, they are called compounds. Most elements are usually found in the combined form. At times, however, some elements are found free.

Minerals may be composed of such single elements. Most minerals, however, are compounds. Carbon, sulfur, gold, silver, and copper are minerals that are single elements. Carbon is the most unusual element in this group. In one form, carbon is the hardest substance known—the diamond. In another form, carbon is a very soft substance—graphite. At this moment you are probably closer to a form of graphite than to a diamond. Graphite may be found in any lead pencil. Pencil lead, you see, is made of a mixture of clay and graphite. The softer the lead, the greater the amount of graphite it contains. Also graphite is often found in home workshops, where it is used as a lubricant. Here is a helpful hint. Did you ever have trouble with a sticking lock? Often, locks stick because they need a lubricant. Next time you have trouble getting a key into a lock, try this. Rub the lead part of a pencil several times over the key. A number 2 pencil is usually good. If the lock needs lubrication, this method often works. Do you know why?

Most Minerals Are Compounds

You recall that oxygen and silicon are the two most abundant elements in the earth's crust. Both of these elements are found in the largest percentage of minerals. When these two elements are chemically combined they form SiO_2 (silicon dioxide). In mineral collections this compound is labeled quartz. Silicon and oxygen also combine readily with several other elements. In such combinations, they often form minerals that are complicated chemical compounds. One of the most common of these is feldspar. There are several types of feldspar; one of the common forms has the formula $KAlSi_3O_8$ (potassium aluminum silicate).

A mineral compound that is often used in the kitchen is halite. This compound is used to flavor food. Have you guessed what it is? It is ordinary table salt. The chemist often describes it as NaCl or sodium chloride. It is also known as common salt or rock salt.

Since there are so many different ways to combine elements, numerous mineral compounds are formed. Accordingly, more than 2,000 different minerals exist on the earth. So that each mineral can be set apart from the others, various tests have been devised. The tests are divided into two groups—chemical and physical.

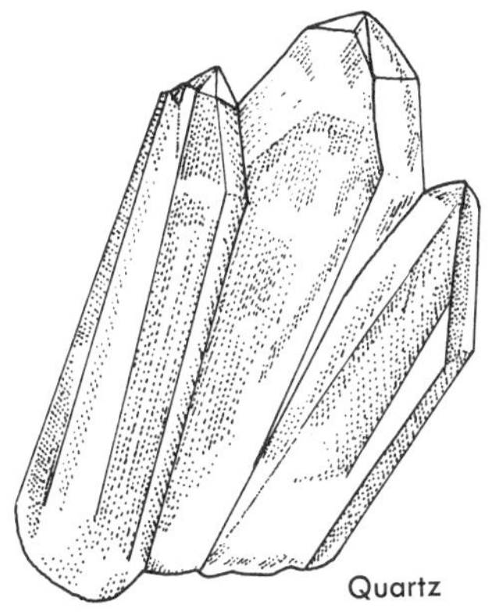

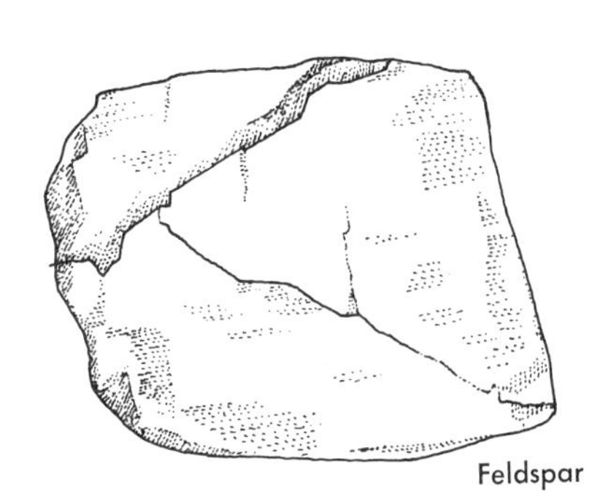

Fig. 14. Rock crystals

Putting Minerals to the Acid Test

A most important factor of minerals is their chemical composition. Because minerals have different chemical compositions, certain chemical tests can be used to identify them.

A simple chemical test is the acid test. This test is especially useful in identifying a group of minerals known as the carbonates. A carbonate is a mineral with CO_3 in its formula. Examples of carbonates are calcite ($CaCO_3$) and siderite ($FeCO_3$). When carbonates are put to the acid test they effervesce, or bubble. The acid test involves placing a drop of hydrochloric acid (HCl) on a mineral sample.

The bubbling that occurs on the surface of carbonate minerals is caused by the release of carbon dioxide (CO_2) gas. The gas is produced by the reaction of the acid with the mineral. A chemist would show the reaction with calcite as follows—

$$2CaCO_3 + 4HCl \rightarrow 2CaCl_2 + 2H_2O + 2CO_2\uparrow$$

There are several other chemical tests, but they are not as simple to do as the acid test. Students who are interested will find them described in books on chemistry or mineralogy.

Physical Tests Are Easy

The first thing most people notice about minerals is their color. Other things readily noticed are the shapes of minerals and the way they shine. A characteristic of minerals that is not usually noticed, however, is hardness. Each of these—color, shine, shape, and hardness—are examples of physical properties. Scientists have developed tests for determining the physical properties of minerals. For some minerals, one test is usually enough to identify them. For most minerals, two or more tests have to be used. Each test gives a clue to the identity of the mineral. Thus, identification of minerals requires good detective work. How would you check out an unknown mineral? You could do the following: first, you could observe its color and its shine; second, you could observe its shape; third, you could determine its hardness; and fourth, you could give it the acid test.

Step No. 1: What Color Is It?

For several minerals, color is a property that does not change. Many minerals, however, can be found in all the colors of the rainbow. Such minerals come in a variety of colors for several reasons, among which may be changes in composition or presence of impurities.

Minerals in which color is an important clue to identity are metallic minerals. Such minerals are the source of many metals. Metallic minerals have what is called a metallic luster (shine). The luster is best observed when the surfaces of such minerals are scratched. You may have seen the same effect when you scraped rust off a piece of metal. Beneath the rust the unchanged metal had a silvery shine. For minerals with metallic luster, color is often an unchanging property. Examples of such minerals are galena and pyrite. Galena, a lead ore, is blue-gray. Pyrite is brass-yellow.

Some minerals without metallic luster may also be found in only one color. Examples are cinnabar, which is red, and malachite, which is green. Cinnabar is a source of mercury; malachite is a source of copper.

Even though most minerals come in several different colors, they can be identified by color. The method used is called the streak test. In this test a mineral is rubbed against an unglazed tile. The mineral usually rubs off against the tile, and the par-

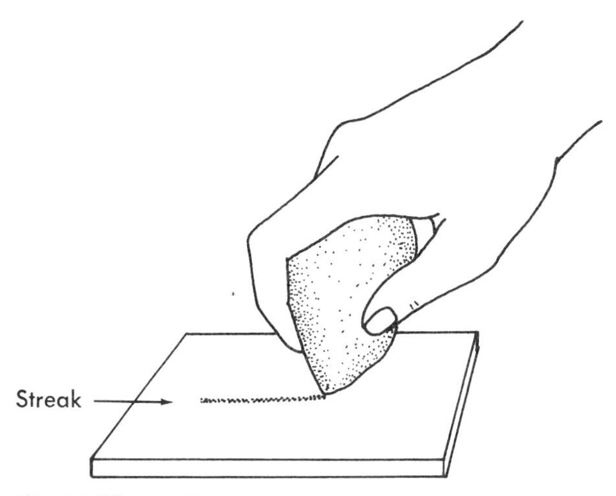

Fig. 15. The streak test

ticles that rub off leave a streak on the tile. It is always the same color, even though the mineral itself may have several colors.

Accordingly, if samples of hematite, pyrite, and talc are subjected to the streak test the following will be observed:

1. The hematite samples always leave a reddish streak.
2. The pyrite samples always leave a greenish or brownish-black streak.
3. The talc samples always leave a white streak.

Each mineral, no matter what its observed color, left the same color of streak, and each mineral has its own characteristic streak.

Step No. 2: What Is Its Shape?

If you could look at the atoms that make up a mineral, you would see that they are arranged in order. The position of the atoms form a "skeleton" for the mineral. Thus, the outside shape of the mineral shows the arrangement of the atoms inside. Scientists call this outside shape the crystal form of the mineral. The atoms in halite form a cubelike "skeleton." The actual sample of halite is shaped like a cube. Thus, the crystal form of halite is a cube. Other minerals having cubic crystals are galena and pyrite.

If you study crystals, you will easily note differences in shape and in the number of sides. For example, the quartz crystal is a 6-sided prism; the diamond crystal, which is 8-sided, looks like two pyramids stuck together at the base; the calcite crystal looks like a slanted box.

An interesting feature of minerals is the way they break. Several minerals, when struck by a hammer or pried apart, tend to break along smooth flat surfaces. The ability of a mineral to split along such plane surfaces is called cleavage. The cleavage planes always follow the flat faces of the crystal forms. Some common minerals that show cleavage are halite, galena, calcite, feldspar, and mica. Halite, galena, and calcite break along three surfaces. Feldspar splits in two directions. Mica splits along one surface only. Not all minerals have cleavage. Many minerals break and form irregular and rough shapes.

Step No. 3: How Hard Is It?

A mineral's scratchability determines its hardness. If you can scratch a mineral with your fingernail, the mineral is softer than your fingernail. Can you scratch a pane of glass with your fingernail? Surely, your answer is no. Glass, then, is harder than your fingernail. What mineral can scratch glass? Since the diamond is the hardest substance known, it scratches glass and almost every other material quite easily. This property of a diamond, plus its beauty, makes it one of the most valuable minerals.

Mineral scientists have established a scale to compare the hardness of different minerals. Look at the scale in Table V. As you can see, the diamond is rated 10 on the hardness scale. Quartz, which is a bit softer, is rated 7. Talc, the softest mineral, is rated 1.

Table V: Scale of Hardness

Hardness	*Mineral*	*Common Substances*
1	Talc	Graphite
2	Gypsum	Fingernail
3	Calcite	Copper Penny
4	Fluorite	
5	Apatite	Knife
6	Feldspar	Glass
7	Quartz	
8	Topaz	
9	Corundum	Emery Cloth
10	Diamond	

Steps No. 4, 5, and 6: When All Else Fails, Try These

The acid test, you recall, helps to pinpoint the carbonate minerals. Try the acid test. This will help you to determine if the mineral sample is a carbonate, such as calcite.

Another test is to observe a mineral's weight. If you picked up a piece of calcite in one hand and a piece of galena in the other you would probably say, "For its size, the galena is heavier." If both pieces were the same size, you would say, "The galena is much heavier." Such comparisons, of course, are only estimates. There is an exact way of comparing the weights of minerals.

This uses the measure called specific gravity. The determination of specific gravity requires special equipment and mathematical calculations. In Table VI the numbers next to the minerals tell you how many times heavier the mineral is than an equal volume of water.

Table VI

Mineral	*Approximate Specific Gravity*
Feldspar	2.5–2.7
Calcite	2.7
Quartz	2.7
Talc	2.8
Chalcopyrite (copper ore)	4.2
Pyrite	5.0
Hematite, Magnetite (iron ores)	5.2
Galena	7.6

Unusual Properties of Minerals

Imagine a mineral that produces electric charges when it is squeezed or when it is heated. Quartz is such a mineral. Other minerals have even more unusual properties. Some minerals such as fluorite, feldspar, and calcite can give off light when rubbed, scratched, or crushed. Fluorite, calcite, and diamonds may give off light rays when exposed to ultraviolet light and X rays.

Some minerals such as uraninite (pitchblende) give off rays without being crushed, scratched, heated, or rubbed. Within the atoms of such minerals certain changes are taking place. As these changes occur, invisible rays and particles are given off. Minerals that do this are called radioactive. Such minerals can be identified by means of a special instrument called the Geiger counter. Geiger counters are often used by prospectors in the search for radioactive minerals.

An unusual, but not uncommon, property of minerals is magnetism. The mineral called lodestone is a natural magnet. Lodestone attracts things made of iron such as nails, paperclips, hairpins, and tacks.

VIII

The Rocks in the Earth's Crust

The First Rocks

Today, as you know, the earth's crust is made of solid rock. Billions of years ago, however, the crust was a bubbling mass of molten materials similar to the lava that pours out of active volcanoes. Today, lava that flows onto the surface cools and forms new rock. In a similar way ages ago, the earth's molten surface cooled and rocks formed. The cooling process must have taken millions of years. Studies indicate that lava flows several feet thick often take years to cool enough to become completely solid. Molten materials thousands of feet thick may take millions of years to become completely solid. Of course, the surface of a molten material solidifies first. In a similar way, a hot pudding cools quickest at the surface and more slowly below the surface. Beneath the relatively cool surface of the earth, we also find hotter materials. At great depths, usually, these materials are molten. Molten materials deep inside the earth are called magma. Magma that reaches the surface of the earth is called lava.

Magma, the Mother Rock

Magma contains all the ingredients needed to make rocks. These ingredients are the elements and compounds we find in minerals. When magma cools these substances are mixed with each other in various ways. Thus, when magma finally becomes solid it contains mixtures of minerals. Such solidified mixtures of minerals are called rocks. Since such rocks were formed from molten materials, they are called igneous rocks. Igneous means "formed by fire."

The Fire Rocks

What would you see if you examined igneous rocks with the naked eye or with a magnifying glass? You would see that some of them are made of crystals of various sizes. You would also note that the crystals are mixed together. These crystals are minerals that formed as the molten materials cooled.

The size of the crystals in a rock tells much about how the rock formed. The faster molten rock cools, the smaller the crystal grains that form.

Fine-grained Rocks

Rocks that form from lava flows usually have very small crystals or no crystals at all. Such rocks are described as being fine-grained or glassy. Examples of such rocks are basalt, felsite, and obsidian.

Basalt, a dark-colored rock, and felsite, a light-colored rock, are fine-grained. The crystal grains in basalt and felsite are too small to be seen with the naked eye. Obsidian has no mineral crystals; therefore, it is as smooth as glass. Samples of obsidian usually look like chunks of smooth black glass. Another common lava rock is pumice. Pumice is filled with many small holes. The holes are formed by gases bubbling through lava as it hardens. Because of the numerous holes, pumice is spongelike and light in weight.

Coarse-grained Rocks

Rocks that form from magma have crystals that vary tremendously in size. Small crystals form in magma that are near the surface and cool relatively fast. Giant crystals form in magma that cools very slowly deep within the crust. Some of the largest crystals ever found were feldspar crystals that weighed over 2,000 tons each.

Rocks that form from magma are usually easy to recognize because of the large crystals they contain. The large crystals give these rocks a coarse-grained appearance. Examples of such rocks are granite, pegmatite, and gabbro.

Granite is the commonest of all igneous rocks. Granites are easy to recognize because the mineral crystals within them are relatively large. In a sample of granite, two different mineral crystals are usually noticeable. One of these crystals, which looks like chunks of cloudy glass, is quartz. The other crystal can be seen to have smooth cleavage surfaces. It is commonly pink, gray, or white. It is the mineral feldspar. Scattered among the quartz and feldspar crystals is a third mineral, which may appear as small dark flakes. If you can scrape small bits out with your fingernail, the mineral is mica. These same minerals are also found in most pegmatites. Pegmatites, however, have much larger mineral crystals in their masses. In fact, the giant feldspar crystals described above were removed from pegmatite rocks.

Gabbro is the coarse-grained equivalent of basalt. A sample of gabbro would appear as a mass of dark-colored mineral grains. A large part of gabbro rock is composed of dark-colored feldspar crystals. Gabbro does not contain quartz.

Rocks from Rocks

At times, muddy water runs in the gutter or in a stream during a rainstorm. The mud in the water is usually made up of tiny rock particles called silt. Do you know where these tiny rock particles come from? They come from the rocks that make up the earth's crust. These particles are usually broken loose from large rock masses by the action of wind, rain, and ice.

What happens to the mud particles being carried in a stream? In time, the stream empties into a relatively still body of water and the mud particles settle to the bottom. Other rock particles are also transported by streams. Such particles are sand, pebbles, and boulders. These particles settle out much more quickly than mud particles, because they are heavier.

Particles that settle out are called sediments. As sediments pile up, the particles on the bottom are squeezed closer together. Dissolved minerals work their way into the spaces between the individual particles. After thousands of years, the dissolved minerals become solid once more. The hardened minerals act as cements and tie the sediments together. In this way, the loose sediments become solid masses of rock. Since such rocks are formed from sediments, they are called sedimentary rocks. The kind of sedimentary rock that forms depends on the sediments and the cements that bind them together.

Sedimentary Rocks Made of Rock Fragments

The types of rock fragments that commonly form sedimentary rocks are particles of silt (muds and clays), grains of sand, and mixtures of sand, pebbles, and boulders. These rock particles range in size from tiny dust-sized specks of silt to huge house-sized boulders.

Rocks from Mud—Shale

When silts are pressed firmly together they harden and form mudstone. Because mudstones are made up of very fine particles they are quite smooth. A very common form of mudstone is shale. Shales are rather soft and crumble quite easily. Colors common in shales are gray and black.

Rocks from Sand—Sandstone

Have you ever heard of Monument Valley? Monument Valley makes up about 2,000 square miles of the Navajo Indian

Fig. 16. Monument Valley, Arizona

Reservation in Arizona. Monument Valley is noted for the thousands of skyscraper-high pillars of sandstone that stand upright on the valley floor. A number of these sandstone pillars are about 1,000 feet high. At one time the whole region was covered by a solid mass of thick sandstone. Over millions of years wind and water carved deeply into the sandstone. The sandstone that remains in the pillars was formed underwater about 25 million years ago. Scientists know that at that time a shallow sea covered large parts of Arizona, Utah, and New Mexico. Huge amounts of sand piled up at the bottom of the shallow sea. In time, the thick beds of sand hardened, because large amounts of dissolved minerals worked their way in between the grains. When these minerals hardened, they cemented the particles together. These natural cements give the sandstones of Monument Valley much of their beautiful red color, which is caused by iron compounds. Cements do not fill all the openings among the sand grains in sandstones. Accordingly, sandstones have many open spaces in them. For this reason, water and other liquids soak through sandstones quite easily.

Rocks from Pebbles and Boulders—Conglomerates

Do you know what conglomerate means? It means made up of different parts. Rocks that are made up of different-sized rock fragments are called conglomerates. If you take a close look at a sample of conglomerate, you will see that it is composed of many pebblelike rocks that are held together by fine material. The appearance of the rock tells you how it was formed. Rock fragments of various sizes such as pebbles and gravel were deposited. Finer materials such as sands filled in the empty spaces among the pebbles and gravel. Dissolved minerals collected in many of the remaining empty spaces. The hardened minerals acted as a cement that held the pebbles and gravel in place.

Rocks from Chemicals—Limestone and Halite

Have you ever visited a natural cave? If you have, you may have been to Howe Caverns in New York, Carlsbad Caverns in New Mexico, Luray Caverns in Virginia, or Mammoth Cave in Kentucky. These are some of the better-known natural caves of the United States. In such caves, you may see chemical rocks

Fig. 17. Luray Caverns, Virginia

forming right before your eyes. Many of the various-sized columns of rock in such caves are growing slightly larger each day by the addition of dissolved rocks being carried by drops of water. Each drop of water gives up some of its dissolved rock materials as it passes over the large columns of rock.

The rock in the caves is usually limestone, which is a form of calcite. You recall how carbonate minerals fizz when acid is added to them. Calcite ($CaCO_3$) is a carbonate mineral. Thus, limestone or calcite will react with acids. In nature, water often acts as a natural acid, breaking down rocks and removing the minerals in dissolved form. The scientist says that the dissolved minerals are being carried away in solution. When such substances in solution become solid again they form chemical sedimentary rocks such as limestone and rock salt (halite).

Substances in solution may settle out under various conditions. Some substances in solution settle out when water evaporates. Accordingly, if a sea dries up the substances in solution will settle out and remain as solid masses.

In limestone caves, the evaporation of water causes the minerals in solution to settle out. Minerals may also settle out when the water is cooled or when the addition of dissolved substances overloads a solution.

Rocks from Animals

In the sea live countless sea animals with shell-like skeletons. Each of these animals is about the size of a speck of dust. During their lifetime they remove minerals from the water to build their shells. When these animals die their shells pile up on the sea floor. After several thousand years, the piled-up shells may form a layer a few inches thick. If the shells pile up for millions of years they can easily form a layer hundreds of feet thick. In time, all the shells may be covered by thick masses of sand or silt. The great pressure of the overlying material would cause the shells to be squeezed tightly together. In this way a thick layer of rock would be formed. There is solid evidence today that the process described above also took place many million years ago. One example of such evidence is the White Cliffs of Dover. These cliffs along the English Channel in Europe tower high above the ships passing by. If some of the rock is powdered and placed under a microscope, tiny shells can be seen. Close examination of these shells reveals that they are the remains of tiny sea animals. The shells of these animals are made of chalk, which is a form of calcite. Thus, the White Cliffs of Dover and other chalk deposits represent the piled-up skeletons of countless billions of sea animals that died millions of years ago.

Changed Rocks

Rocks that have already formed may be changed into other forms of rocks. Changes are usually caused by great heat, high pressure, or chemical action. Some changes that occur in rocks can be in hardness, crystal form, and chemical composition. Changed rocks usually look and react differently from the rocks that they came from. Accordingly, changed rocks are called metamorphic rocks. Metamorphic means "changed in form."

Common Metamorphic Rocks

If you examine a small piece of slate, you will note that it looks somewhat like shale. Slate is metamorphosed shale. You will also find that the slate is harder than shale and far less likely to crumble. Slates have a special kind of cleavage that makes it easy to split them into large flat slabs.

When sandstone is metamorphosed, the openings between the grains of sand disappear. Thus, the resulting metamorphic rock, quartzite, does not absorb water.

The world's most beautiful rocks come from limestones. When limestones are metamorphosed they become harder. In addition, rounded grains form in the masses of rock. The changed crystal-like rock is called marble. Pure white marble looks somewhat like a solid mass of sugar. Often marbles contain various colored streaks. The streaks are caused by minerals that formed in the original limestones during metamorphism.

When granite is metamorphosed the minerals become squashed and they spread out. As a result, metamorphosed granite has a banded appearance. Such metamorphic rocks are called gneiss. Gneiss may also form from sedimentary rocks such as conglomerate. Under great heat and pressure the pebbles in conglomerates spread out and form bands or layers.

Table VII: Sources of Some Sedimentary and Metamorphic Rocks

Sediment	*Sedimentary Rock*	*Metamorphic Rock*
clay	shale	slate
sand	sandstone	quartzite
pebbles and sand	conglomerate	gneiss
calcite	limestone	marble

IX

Earth—Our Storehouse

Rocks and Minerals All Around Us

Look around your home, your neighborhood, your school. Almost everything you see is made either partly or fully from rocks or minerals. If you made a list of the things that come from rocks and minerals it would include highways, bridges, cars, sidewalks, stoops, windows, walls, wires, lights, radios, clocks, pencils, coins, pans, dishes, jewelry, cosmetics, medicines, paints, and so on.

Rocks and Minerals for Building

Did you ever hear of a sand mine? All over the United States millions of tons of sand are removed from sand mines regularly. These sands are sediments that were broken loose from the earth's solid crust. Such sands are used in the making of concrete, which goes into the construction of highways, runways, sidewalks, buildings, bridges, dams, and pipes.

A familiar sight on the highways are trucks transporting concrete to construction sites. Sometimes the concrete is mixed in the truck as it travels to the site. The concrete mix includes sand, gravel or other small rock fragments, cement, and water. After the concrete mix is poured, it hardens and forms a man-made stone somewhat like conglomerate. You recall that conglomerate is a sedimentary rock of sand and pebbles held together by natural cements. The cement in concrete is made from sedimentary rocks such as limestone and shale.

How Is Cement Made?

The cement that is used in concrete is usually portland cement. It is largely made of limestone and shale (or clay). These materials are crushed and mixed with water. Then the mixture is rotated slowly in a large container and heated at about 270 degrees F. This process changes the mixture into marble-sized balls. The balls, called clinkers, are crushed into a fine powder (generally gray-colored) called portland cement. You may have seen portland cement in paper sacks in lumber yards, hardware stores, or around construction sites.

Cement, the Construction Glue

When primitive man learned to build shelters of stone, he soon found that he needed something to hold the stones together. He probably used mud and clay as cements. Unfortunately, these natural cements were weak and were easily washed away by rain. Thus, in places where rainfall occurred quite regularly, stone shelters probably fell apart rather easily. About 6,000 years ago the Egyptians found a much better way to cement stones together. They discovered that lime and gypsum mixed together made a strong cement, and they used this cement in building the great pyramids.

Modern cement is vital to the construction industry. You recall how cement is used in the making of concrete. Cement is also used to hold bricks, stones, and concrete blocks together. When used in this way it is mixed with sand and water and made into a pastelike mass. This mass is called mortar. In construc-

Fig. 18. Colonial brick building

tion, a thin layer of mortar is placed between building stones such as brick or concrete blocks. When the mortar hardens it becomes hard as stone and cements the building stones tightly into place.

Bricks and Stones May Make New Homes

Long before recorded history, people probably used blocks of sun-baked clay as bricks. As civilization grew, man-made bricks of clay were produced. These bricks were used in construction by ancient civilizations such as those of India, China, Egypt, Greece, and Rome. The art of brickmaking and brick-building was spread across Europe by the Romans. In America, in the 1600's, the early English colonists built houses of brick. To this day, bricks are widely used by the construction industry. Brick-making factories make hundreds of millions of bricks each year.

Brickmaking involves three basic steps. Clay, which forms in nature from the mineral feldspar, is powdered and mixed with water. The mixture forms a stiff plasticlike mud. This material is shaped into bricks, and the "wet bricks" are then baked in a furnace, where they harden.

Another important building "brick" is the concrete block. Concrete blocks are simple to make. A concrete mix is poured into a mold. When the mix hardens the mold is removed. The finished block is somewhat bigger than a large shoe box. Concrete blocks are commonly used in constructing the foundations and the walls of buildings.

Natural Building Blocks

Building stones that come directly from bedrock are usually granite, marble, limestone, and sandstones. These rocks are cut in large blocks from bedrock in stone quarries. In many quarries huge saws are used to cut these large slabs from the bedrock.

Slabs of granite are often used as steps of buildings because granite is hard and long-lasting. Polished granite, marble, and limestone are commonly used for covering the inside and outside walls of public buildings. Both marble and granite are also widely used in making monuments.

In past years, large amounts of sandstone were used for building homes. In many Eastern cities, block after block of "brownstone houses" can be seen. The brownstone is actually a dark brown sandstone.

More Building Materials

Millions of homes have plaster walls. Plaster is made from the mineral gypsum. When gypsum is heated, it forms a powdered substance called plaster of Paris. When mixed with water, plaster of Paris forms a thick white paste called plaster. The wet plaster can be smoothed onto walls or ceilings, where it sets.

Large amounts of plaster are used to make plaster board. These large flat boards consist of a layer of plaster about ½ inch thick, sandwiched between heavy paper. Plaster board has largely replaced plaster for finishing the inside walls of homes.

Skeletons of Steel

Steel makes the skeletons or frames of many products used in the home and industry. Each year millions of tons of steel go into the manufacture of automobiles, trains, ships, refrigerators, desks, "tin" cans, and toys. In addition, most large buildings and bridges are made of steel. Steel girders used for buildings form strong skeletons to which floors and walls are attached. In buildings made of concrete, steel rods are locked into the concrete for added strength. After most buildings are completed, the steel is hidden from view. By comparison, steel bridges appear as skeletons of "naked steel."

The major ingredient of steel is iron. Most iron in the United States comes from rich deposits of ore around Lake Superior and near Birmingham, Alabama. These deposits, composed of the mineral hematite (Fe_2O_3), are a compound of iron and oxygen. The steps involved in making an iron ore, such as hematite, into steel include separating the iron from its compound. Essentially, this is done by placing a mix of iron ore, limestone, and coke (coal rich in carbon) in a furnace. The mix is melted, and air is blown through it. This process releases the oxygen from the iron ore to form a relatively pure form of iron. This iron, which is in the hot liquid form, is emptied from the bottom of the furnace and is cast into ingots called pig iron. Pig iron is made into different kinds of steel by the addition of carbon or metals such as manganese or chromium.

Copper and Aluminum—The Old and the New

Copper and aluminum are among the most widely used metals. These metals are especially important in the transportation, communication, and building industries. Although both metals have similar uses, copper has been around longer. Ages before man learned about iron, he knew about copper. Thousands of years ago men softened copper in fires and shaped it into such objects as tools and utensils. By comparison, the use of aluminum is in its infant stage. The first time aluminum was separated from its ore in quantity was in 1888.

Copper and aluminum are used in large amounts by the building industries, in the making of household appliances, and in producing electric wire and cables. Aluminum has been found to be very good for outside walls, window frames, screens, and rain gutters.

Copper may be mined as native copper or it may come from minerals such as azurite and cuprite. Mines are located in many of the Western states. Separating copper from its ore involves three basic steps: (1) melting the ore; (2) separating the impurities; and (3) forcing air through the molten mass to further purify the copper. Copper may be used in the pure form or to make alloys that are mixtures of metals. One of the best-known alloys of copper is brass.

Aluminum is obtained from bauxite. The major source of bauxite in the United States is in Bauxite, Arkansas. Separating aluminum from its ore involves several steps, which result in the formation of a fine white powder. The powder is changed into pure molten aluminum in huge electric furnaces.

The Finishing Touches

How often have you marveled over the smooth finish on a marble slab or a piece of furniture? Minerals commonly used to smooth surfaces are called abrasives. Common abrasives found around the home are sandpaper, emery boards, pumice stones, and scouring powder.

Ordinary sandpaper is made by gluing sand grains (quartz) to paper. Sand is also found in harsh scouring powders. A special kind of scouring powder is toothpaste; an abrasive commonly used in toothpaste is made of a light rock that formed as a deposit of the shells of tiny sea plants. Emery boards are made from corundum, which is number 9 on the hardness scale, second only to the diamond. Grinding wheels are made of corundum. Pumice is a lightweight rock found near volcanoes. Pumice stones are used to rub away calluses on the body. Pumice in the powdered form is rubbed on furniture to achieve a high polish.

Sunlight Makes It Go

In recent years, experimental cars have been made to run on sunlight. One experimental car has a roof made of special metal plates that produce electricity in sunlight. The electricity makes a motor go, and the motor makes the wheels turn. This type of car is just fine—until the sun goes down. Without sunlight, it does not run. Many more experiments will be needed to figure out how to run such a car at night. One possibility is to use gasoline to run the car at night. Gasoline is a liquid fuel that contains large amounts of carbon and hydrogen. When gasoline is burned the carbon and hydrogen combine with oxygen. In a gasoline engine, the burning fuel produces energy, which can do work. Yet, when we use gasoline we are still using sunlight. Sounds strange, does it not? What does sunlight have to do with gasoline? The answer is simple. Gasoline is a man-made chemical that comes from petroleum. Petroleum is a natural oil usually found trapped in the earth's crust. Most of us call petroleum, oil or crude oil.

Oil formed in the ground millions of years ago from the remains of very small plants and animals. When these organisms were alive they needed sunlight to grow. The plants used sunlight to make food, and the animals lived on the plants and other life forms. Sunlight made life possible for all the different life forms on earth. When the plants and animals died, their remains piled up in shallow water and in time were covered with sediments. Over many years, the buried plant and animal remains underwent chemical changes that transformed them into oil. The oil seeped into openings in rocks and was trapped. Today, oil workers drill into these rocks to remove the oil. Part of this oil, when refined, is made into gasoline. Thus, when we burn gasoline, we are using the energy of sunlight to make things work.

Where Do We Find Oil?

Oil may be found in many places all over the world. It may be found above the Arctic Circle, at the South Pole, in the Sahara, in the ocean, and in mountainous regions. In the United States, oil wells are operating in about half of the states. Most of this oil is found in Texas, Louisiana, Oklahoma, and California.

Much of the world supply of petroleum is trapped deep beneath the surface. Geologists locate oil by studying rock formations and fossils in the rocks. Wherever they find the right kinds of rocks and fossils, they drill wells.

How Do We Get Oil Out of the Ground?

To get oil out of the ground, we must drill holes into the crust. Often the drill holes must go through solid rock to reach reservoirs of oil. Many drill holes go down more than a mile be-

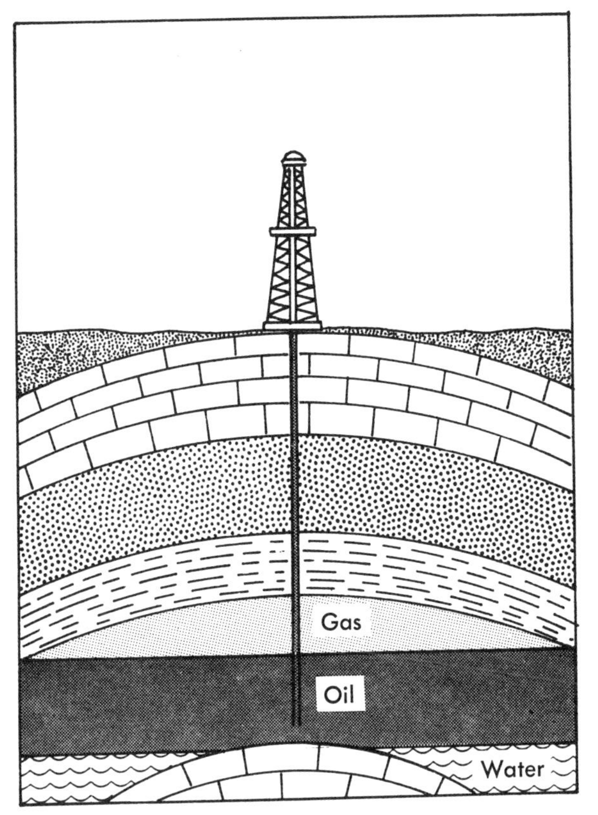

Fig. 19. Drilling for oil

fore oil is reached. By comparison, the first hole drilled in this country, in Pennsylvania in 1859, reached oil about 69 feet below the surface.

Places in the crust where oil has collected are called traps. In a typical trap, oil collects in rocks such as sandstone and limestone. These rocks can absorb oil because they contain many open spaces or pores. At or near the surface, these open spaces are commonly filled with air. At times, oil seeps into the pores and pushes out the air. The oil becomes trapped when it meets layers of rock such as shale. Shale does not let oil (or other liquids) through because its pores are too small.

Often the rock layers are curved upward like a dome. Gas, oil, and water collect in such domes. When a dome is drilled into, the gas is first to rush out. Then the oil follows. Can you explain why the gas comes out first?

The natural gas that comes out of oil wells is not wasted. It is useful as a fuel. It is sent to homes and to factories through pipelines. The gas may be used for cooking, heating, and for the production of electricity. Many people prefer to use gas as a fuel rather than coal and oil: gas is easier to ship, it leaves no ashes, and it burns cleaner.

How Do We Get Gasoline from Oil?

The crude oil that comes out of the ground is thick and sticky. It has to be refined to make gasoline from it. In the refining process, the crude oil is heated. The vapors that boil off are cooled, and the cooled vapors condense into a liquid. This liquid is gasoline. After all the gasoline is boiled off, the temperature is raised. At a higher temperature the crude oil gives off kerosene vapors. The kerosene vapors are cooled and the liquid is removed. Each time the temperature of the crude oil is raised, another substance boils off. Thus, several other substances can be separated from crude oil. Among these are fuel oils, lubricating oils and greases, and various waxes.

Sunlight in Coal

Just as oil had its origin in sunlight, so did coal. Coal, however, formed from plants somewhat like the ferns that grow today; but these fernlike plants were as big as trees. These ancient trees thrived in swamps millions of years ago. Their leaves took in sunlight and absorbed carbon dioxide from the air. The roots absorbed water from the ground. Within the leaves, the water and carbon dioxide were changed into plant materials. In this way, the trunks, branches, and leaves stored large amounts of carbon and hydrogen. As these trees died, they piled up in the swamps, and over millions of years sediments piled up on top of the dead trees. In time, the trees were buried deep below sediments. The heavy sediments pressed hard on the plant parts, causing them to decay and form brown spongy masses. Such partly decayed plant parts are called peat. In some places, peat is used as a fuel. With the passage of time, peat becomes packed together more tightly. Chemical changes remove gases and liquids from the peat; it dries out, hardens, and thus becomes more solid. In this form it is called lignite. Lignite contains more carbon than peat, which makes lignite a better fuel. However, lignite is considered a poor grade of coal. Little lignite is used in this country, because we have available much larger deposits of higher-grade coal. One of these higher-grade coals is bituminous.

Bituminous coal forms from lignite that has been under great pressure for a long time and thus has become more solid. Much hydrogen is driven out. Thus, the mass becomes richer in carbon, forming bituminous coal. Bituminous is commonly a dull black in color and breaks rather easily. Often called "soft coal," it is the most commonly used coal in the world.

The highest grade of coal is anthracite, or "hard coal." Anthracite is nearly pure carbon. It forms from bituminous coal in regions where mountain-building forces are at work. Mountain-building forces produce great heat and great pressures. Under such heat and pressures, bituminous coal is further changed and hardened. The resulting coal—anthracite—is shiny, hard, and black like tar. Anthracite is highly valued as a fuel for it burns hot and clean and lasts for a long time.

Anthracite coal may be further changed by greater heat and

pressures. Then the atoms within the hard coal take new positions and hydrogen atoms are forced out. The new substance formed is graphite, which is pure carbon. And strange changes have taken place. Graphite is a very soft substance. It is so soft that it can be scratched easily with a fingernail. Stranger yet, this soft mass of carbon cannot be used as a fuel. The reason is simple—it does not burn at ordinary temperatures. It burns at a temperature of about 7200 degrees F. By comparison, coal burns at temperatures below 1000 degrees F. When powdered, graphite may be used as a lubricant. Ordinarily, it is sold in tubes at hardware stores. It is especially useful in outdoor locks such as those in automobiles. In extremely cold weather, oil may jam a lock.

Although coal, oil, and gasoline provide most of our heat and energy, we are beginning to make more use of nuclear energy. Nuclear energy comes from the splitting of uranium atoms. Uranium commonly comes from pitchblende. Large pitchblende deposits are found in several of the Western states.

Sunshine Makes Diamonds Too!

Imagine a diamond forming from a tree trunk. Do you think this is possible? It is—because diamonds are a pure form of carbon. The steps in the formation of diamonds are much the same as with coal. You recall, in the presence of sunshine, plant leaves manufacture plant materials. The plant materials contain large amounts of carbon and hydrogen. Under pressure and heat, the plant parts become coal. And, under greater heat and pressure, graphite is formed. Diamonds are believed to form under similar conditions. They form from carbon deposits such as coal or graphite. The carbon deposits probably changed to diamonds as a result of volcanic action. In fact, diamonds are found in places where there were once ancient volcanoes. Scientists believe that deep in the crust molten rocks moved upward through openings in volcanoes. The molten rocks came in contact with carbon deposits, and under such great heat and pressure the carbon atoms took new positions. In their new positions, they formed pyramidlike shapes. Today, we find them in this shape. In this form, it is very difficult to tear the carbon atoms apart. Thus, diamonds are very hard—in fact, the hardest substance known to man. Their hardness makes them very useful. They are used to cut through steel, rock, and other hard substances. And of course, when polished, diamonds are treasured for their beauty and brilliance.

Minerals Have 1001 Uses

In this chapter we have seen a number of ways in which rocks and minerals are valuable. In Table VIII are listed many other ways in which rocks and minerals are used by man.

Table VIII: Some Common Rocks and Minerals and Their Uses

Azurite	Copper (wire, nails, utensils, coins, screens)
Basalt	Crushed stone (filler in highways and railroad beds)
Bauxite	Aluminum (airplanes, automobiles, trains, wire, nails, utensils, screens, window frames)
Calcite	Building material (travertine, "marble"); making steel, cement, limestone
Chalcocite	*See* Azurite
Cinnabar	Mercury (thermometers, barometers, dental work, mirrors)
Chrysotile	Asbestos (brake linings, shingles, insulation)
Clay	Brick; tile; china; pottery
Corundum	Abrasive (emery); as a gem stone (ruby, sapphire)
Diamond	Grinding and polishing other diamonds, rocks, and hard substances; as a gem stone
Galena	Lead (plumbing, type for printing, storage batteries, paints)
Garnet	Abrasive; as a gem stone
Granite	Building stone; curbstone; paving blocks (Belgium blocks); monuments
Graphite	Dry lubricant
Gypsum	Plaster of Paris (wallboard, walls, ceilings, statues)
Halite	Table salt; rock salt for melting ice on highways

Hematite	Iron (steel for buildings, bridges, highways, railroads, autos, appliances)
Kaolinite	*See* Clay
Limestone	*See* Calcite; agriculture (to treat acid soils)
Magnetite	*See* Hematite
Marble	Building stone; monuments; statues
Mica	Heat insulation (vermiculite); electrical insulation (toasters, irons)
Pegmatite	Source of mica and gem minerals
Pitchblende	Uranium (atomic reactors, atomic weapons)
Pyrite	Source of sulfur (sulfuric acid, medicine, fertilizers, rubber, insecticides, explosives)
Quartz	Abrasive (sandpaper); glass (windows, glasses, etc.)
Shale	Bricks; clay; cement
Slate	Roof tiles; chalk boards
Uraninite	*See* Pitchblende